Math Mammoth Grade 5 Tests and Cumulative Reviews

for the complete curriculum
(Light Blue Series)

Includes consumable student copies of:

- Chapter Tests
- End-of-year Test
- Cumulative Reviews
- Fraction Cutouts

By Maria Miller

Contents

Chapter 1 Test

1. Solve (without a calculator).

   a. $1,456 \div 26$

   b. $18,755 \div 31$

   c. 391×475

2. Solve: $Y - 8,687 = 19,764$

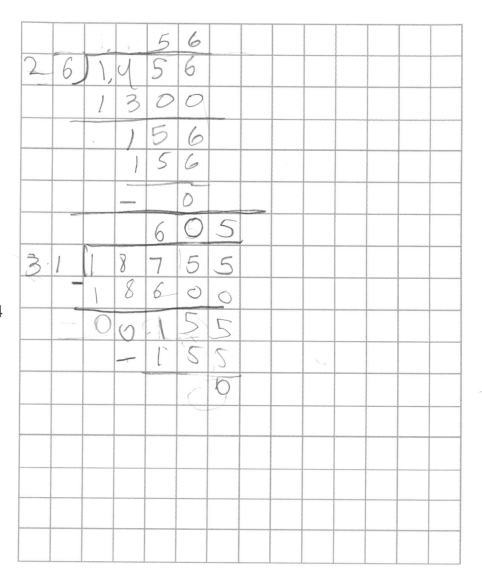

```
            5  6
    26 ) 1, 4  5  6
         1  3  0  0
            1  5  6
            1  5  6
            -     0
               6 (0) 5

    31 ) 1  8  7  5  5
         1  8  6  0  0
       (0) 0 - 1  5  5
         -  1  5  5
                  (0)
```

3. Solve in the right order.

a. $2 \times (80 - 8) = \underline{144}$	**b.** $100 - (240 \div (8 + 2)) = \underline{76}$ 10

4. Divide mentally. $9\overline{)8,109}$

a. $\dfrac{8,109}{9} = 901$	**b.** $\dfrac{1,244}{4} = 311$	**c.** $\dfrac{4,045}{4+1} = 809$

5. Find a number to fit in the box so the equation is true.

a. $42 = (\boxed{31} - 10) \times 2$	**b.** $48 \times 10 = \boxed{80} \times 6$ 480

Product
Quotient

6. Write an expression *or* an equation to match each written sentence. You do not have to solve anything.

a. The product of *s* and 11	**b.** The quotient of 48 and *b* is equal to 8.
$2+11=13$	easy ? but forgot what a quotient is

7. Write a <u>single</u> expression (number sentence) for the problem, and solve.

Mia bought 5 pairs of socks for $2.50 each, and paid with a $20 bill. What was her change?

$\$20 - \$12.50 = \$7.50$

8. Draw a bar model to represent each equation. Then solve them.

a. $5 \times Y = 600$	**b.** $Z \div 3 = 140$
 $5 \times$ 600 120 5× 120 600 Y	 $Z = 140 \times 3$ $Z = 420$ $2+2+2 = 6 \rightarrow 2 \times 3 = 6$ $6 \div 3 = 2$ $6 \div 2 = 3$

9. Is 991 divisible by 3?

Why or why not?

10. Factor the following numbers to their prime factors.

a. 16	**b.** 34	**c.** 80
/\ 8×2 4 2 2 2 2	/\ 17×2 1 2	/\ 8×10 2 4 2×5 2 2

6

Chapter 2 Test

The calculator is not allowed in the first six problems of the test.

1. Write the numbers.

   a. 70 million 6 thousand 324   070,606,324,000

   b. 4 billion 32 thousand   4,000,032,000,000

   c. 98 billion 89 million 98   098,089,000,098

2. What is the *value* of the underlined digit?

a. 410,2**9**3,004	b. 408,0**3**7,443,000	c. **4**,395,490,493
Value: 90,000	Value: 30,000,000	Value: 4,000,000,000

3. Round these numbers as indicated.

number	183,602	355,079,933	29,928,900
to the nearest 1,000	184,000	355,080,000	29,929,000
to the nearest 10,000			
to the nearest 100,000			
to the nearest million			

4. Solve.

a. $9^2 =$ _____	b. $10^3 =$ _____	c. $3^3 =$ _____

5. Write using exponents, and solve.

a. six squared =	b. two to the fifth power =

6. Calculate the products mentally.

a. $40 \times 900,000$	b. $600 \times 200 \times 500$
c. 7×10^4	d. 48×10^6

7. Complete the math path. (Calculator usage is optional.)

4 million	subtract → 700 thousand		add → 12 million	

add ↓ 3 billion

	subtract ← 20 thousand		add ← 8 hundred	

8. First, estimate the answer. Then calculate the exact answer and the error of estimation using a calculator.

a. $209,800 - 4,730$

Estimation: _____

_____

Exact answer: _____

Error of estimation: _____

b. $2,543 \times 5,187$

Estimation: _____

_____

Exact answer: _____

Error of estimation: _____

c. $56,493,836 + 345,399 + 7,089,400$

Estimation: _____

Exact answer: _____   Error of estimation: _____

9. In October 2008, the US congress approved a $700 billion "bailout plan" to aid the failing banks. There were about 305,000,000 persons in the USA. If the cost of this bailout was divided evenly between all the U.S. people, how much would a family of four in the USA have to pay for it? Give your answer rounded to the nearest hundred dollars.

Chapter 3 Test

1. Write an equation to match the balance. Then solve what *x* stands for.
   Remember to write 2x to mean 2 x's in the same pan, and 3x to mean x, x, and x in the same pan.

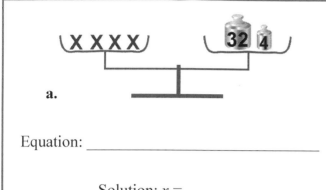

a.

Equation: _____

Solution: *x* = _____

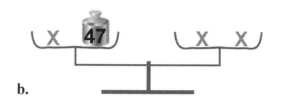

b.

Equation: _____

Solution: *x* = _____

2. Write an equation for each bar model. Then, solve for x.

x	*x*	100
← —— 302 —— →

a.

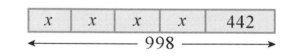
← —— 998 —— →

b.

3. A cell phone that costs $48 is on sale with 1/6 off of the normal price.
   How much would *three* discounted phones cost?

4. Two sisters divided 250 smooth beach rocks so that
   the elder sister had 32 rocks more than the younger sister.
   How many rocks did the younger sister get?

5. Five kilograms of potatoes costs $7.50. Henry bought 2 kg.

   a. How much do 2 kg of potatoes cost?

   b. What is Henry's change from $10?

6. A high-quality hard drive costs three times as much
   as a low-quality one. Buying the two together would cost $820.
   How much does the low-quality hard drive cost?

7. Matthew is 3/8 as tall as his dad.
   If Matthew is 66 cm tall, then how tall is his dad?

Chapter 4 Test

As this test is quite long, feel free to administer it in two parts.

1. Write the decimals indicated by the arrows.

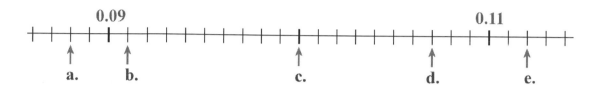

a. _____ b. _____ c. _____ d. _____ e. _____

2. Complete the addition sentences.

   a. $1.3 + \underline{} = 7$ **b.** $0.76 + \underline{} = 1$ **c.** $3.65 + \underline{} = 4$ **d.** $0.18 + \underline{} = 0.2$

3. Write as decimals.

   a. $\dfrac{21}{100} =$ **b.** $\dfrac{46}{1000} =$ **c.** $3\dfrac{7}{100} =$ **d.** $20\dfrac{2}{10} =$

4. Write as fractions or mixed numbers.

   a. 0.6   **b.** 0.82   **c.** 1.208   **d.** 0.093

5. Compare.

   a. $0.05 \ \square \ 0.2$ **b.** $0.43 \ \square \ 0.045$ **c.** $2.05 \ \square \ 2.051$ **d.** $0.438 \ \square \ \dfrac{1}{2}$

6. Round the numbers to the nearest one, nearest tenth, and nearest hundredth.

rounded to...	nearest one	nearest tenth	nearest hundredth
8.816			
1.495			

rounded to...	nearest one	nearest tenth	nearest hundredth
0.398			
9.035			

7. Solve.

a. $0.4 \times 7 =$ _____	**c.** $20 \times 0.05 =$ _____	**e.** $0.2 \times 1{,}000 =$ _____
b. $7 \times 0.09 =$ _____	**d.** $100 \times 0.09 =$ _____	**f.** $0.8 \times 0.8 =$ _____

8. Divide.

a. $0.24 \div 6 =$ _____	**c.** $2 \div 100 =$ _____	**e.** $0.43 \div 10 =$ _____
b. $0.081 \div 9 =$ _____	**d.** $0.8 \div 10 =$ _____	**f.** $7 \div 1000 =$ _____

9. Multiply and divide using powers of ten.

a. $0.05 \times 10^4 =$ _____	**c.** $3.5 \div 10^2 =$ _____
b. $10^5 \times 7.8 =$ _____	**d.** $13{,}200 \div 10^4 =$ _____

10. Find the number that is 1 tenth and 2 thousandths more than 1.109.

11. **a.** Estimate the answer to 0.6×21.8.

    b. Now find the exact answer to 0.6×21.8.

12. Divide using long division:

    a. $7.836 \div 6$

    b. $21 \div 4$

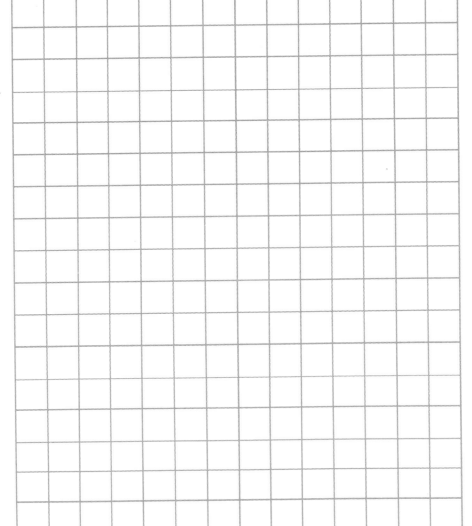

13. Is the answer to 0.9×0.8 more or less than 0.8?

　　Explain in your own words why that is so.

　　Is it more or less than 0.9?

14. Teresa packed 7 kg of blueberries equally into four boxes.
　　How much does each box weigh?

15. Convert.

a. 0.7 m = _____ cm	b. 2,650 ml = _____ L	c. 5.16 kg = _____ g
3.2 km = _____ m	0.9 L = _____ ml	400 g = _____ kg

16. Convert.

a. 8 ft 10 in. = _____ in.	b. 2 gal 3 C = _____ C	c. 81 oz = _____ lb _____ oz
183 in. = _____ ft _____ in	45 oz = _____ C _____ oz	165 oz = _____ lb _____ oz

17. Samuel bought a 0.9-liter box of juice and two cans of juice, 350 ml each.
　　What is the total volume of the juice he bought?

18. Mary bought a 2-kg bag of tomatoes for $4.48.
Then, Mary sold 250 g of the tomatoes to her friend.

   a. How much does half a kilogram of tomatoes cost?

   b. How much did she charge her friend?

19. A DVD that normally costs $19.95 is discounted.
The new price is 2/5 off of the normal price.
Find how much *two* copies of the discounted DVD cost.

20. In the spring, Arnold weighed what apples that were still in the root cellar.
He gave half of the apples to his neighbor, and divided the rest equally into four boxes.
Each box weighed 0.47 kg.

   a. Label in the model the parts that equal 0.47 kg.

   b. Label in the model the total weight
   of all the apples with "???"

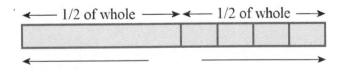

   c. Find the total weight of the apples that Arnold found in the cellar.

Chapter 5 Test

1. Plot the point (9, 5) on the grid. Then, plot the point that is two units down and four units to the left from that point. What are its coordinates?

2. Plot the points from the "number rule" on the coordinate grid. Fill in the rest of the table first, using the rule.

   The rule is: $y = 2x - 1$.

x	1	2	3	4
y				

x	5	6	7	8
y				

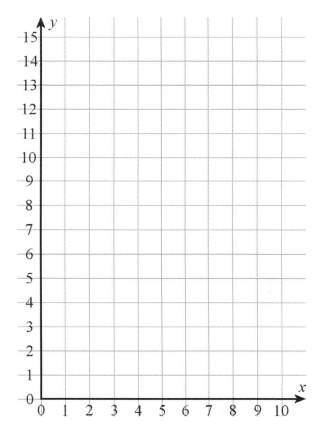

3. A cell phone store kept track of how many cell phones they sold each day of the week ("units sold"). A certain day they started a promotion with 3/10 off of the normal price.

   a. Add the number labels for the vertical axis next to the tick marks (the scaling).

   b. Plot the remaining points and finish the line graph.

   c. Which day did the promotion most likely start?

Day	Units sold
Mo	17
Tu	14
Wd	15
Th	21
Fr	19
Sa	23
Mo	15
Tu	34
Wd	40
Th	37
Fr	33
Sa	41

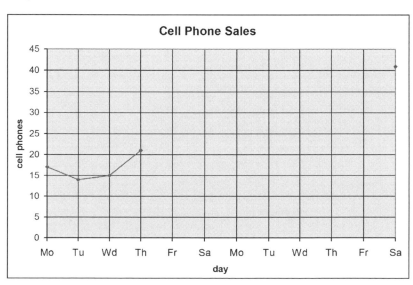

4. Mary asked 20 people in a club how old they were (in years). Here is her data:
   10, 9, 10, 12, 15, 8, 9, 10, 11, 13, 11, 10, 9, 12, 9, 13, 11, 10, 15, 14
   (Each number is the response from one person.)

   a. Fill in the frequency table. Make four categories. Draw a bar graph.

Age	frequency

   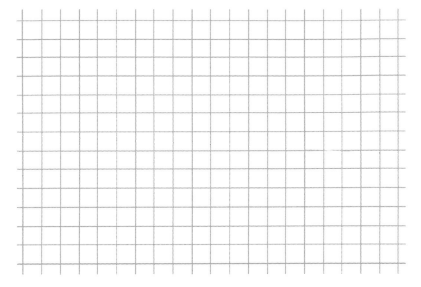

   b. What is the mode of this data set?

   c. Find the average.

5. The chart shows Alice's science test scores for five different tests.

Alice's test scores	
Test 1	76
Test 2	66
Test 3	74
Test 4	81
Test 5	88

   a. Draw a line graph.

   b. Calculate the average.

   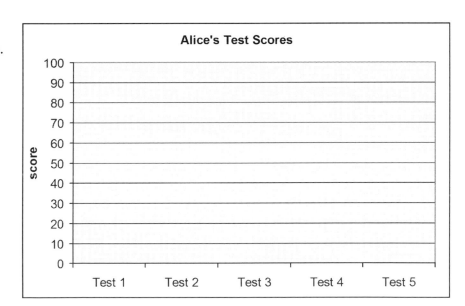

   c. Plot the average in the line graph.

   d. If the test where Alice scored the worst was dismissed and not taken into account, what would Alice's average score be?

Chapter 6 Test

1. Write as mixed numbers.

    a. $\dfrac{26}{3}$ **b.** $\dfrac{45}{7}$ **c.** $\dfrac{34}{5}$

2. Add and subtract.

a. $7\dfrac{6}{8}$ $+ \ 2\dfrac{5}{8}$	**b.** $6\dfrac{1}{5}$ $- \ 3\dfrac{4}{5}$	**c.** $4\dfrac{6}{11}$ $+ \ 9\dfrac{9}{11}$ $+ \ 2\dfrac{4}{11}$

3. Find the missing fractions or mixed numbers.

a. $2\dfrac{3}{7} + \underline{\hphantom{xx}} = 5\dfrac{1}{7}$	**b.** $2\dfrac{5}{9} + 4\dfrac{6}{9} + \underline{\hphantom{xx}} = 10$	**c.** $7\dfrac{2}{15} - \underline{\hphantom{xx}} = 2\dfrac{8}{15}$

4. Mark the fractions on the number line. $\dfrac{2}{3}, \ \dfrac{5}{6}, \ \dfrac{7}{12}, \ \dfrac{3}{4}, \ \dfrac{11}{12}$

5. If you can find an equivalent fraction, write it. If you cannot, cross out the whole problem.

a. $\dfrac{3}{7} = \dfrac{\hphantom{xx}}{21}$	**b.** $\dfrac{4}{3} = \dfrac{\hphantom{xx}}{18}$	**c.** $\dfrac{5}{6} = \dfrac{\hphantom{xx}}{11}$	**d.** $\dfrac{2}{5} = \dfrac{8}{\hphantom{xx}}$	**e.** $\dfrac{5}{6} = \dfrac{15}{\hphantom{xx}}$

6. Compare the fractions, and write < , >, or = in the box.

a. $\dfrac{7}{4} \ \square \ \dfrac{5}{3}$	**b.** $\dfrac{5}{11} \ \square \ \dfrac{1}{2}$	**c.** $\dfrac{7}{10} \ \square \ \dfrac{69}{100}$	**d.** $\dfrac{3}{4} \ \square \ \dfrac{75}{100}$	**e.** $\dfrac{8}{7} \ \square \ \dfrac{7}{9}$

7. Draw something in the picture and explain how we can add 1/3 and 2/5.

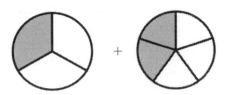

8. Add and subtract.

a. $\dfrac{2}{3} + \dfrac{3}{4}$	**b.** $\dfrac{5}{6} - \dfrac{2}{3}$
c. $3\dfrac{1}{7} - \dfrac{1}{2}$	**d.** $6\dfrac{7}{8} + 3\dfrac{1}{5}$

9. Write the fractions in order starting from the smallest.

$$\dfrac{4}{7}, \dfrac{5}{9}, \dfrac{7}{5}, \dfrac{1}{2}$$

10. Write a fraction addition with a sum (answer) of 1 where one fraction has the denominator of 4, and the other has the denominator of 7.

11. Measure the sides of the triangle in inches. Find its perimeter.

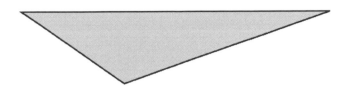

Chapter 7 Test

1. Write the simplification process.

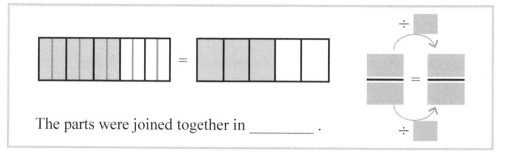

    The parts were joined together in _____ .

2. If possible, simplify the following fractions. Give your answer as a mixed number when possible.

a. $\dfrac{22}{6} =$	b. $\dfrac{28}{42} =$	c. $\dfrac{35}{32} =$

3. Julie needs 2/3 cups of butter for one batch of cookies.
   Find how much butter she would need to make five batches of cookies.

4. Draw a picture to illustrate $5 \times \dfrac{3}{4}$ and solve.

5. Is the following multiplication correct?

   If not, correct it.

   $\dfrac{3}{4} \times$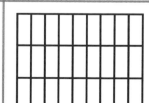

6. Multiply the fractions, and shade a picture to illustrate the multiplication. Simplify your answers.

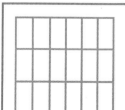 a. $\dfrac{2}{3} \times \dfrac{1}{6}$	b. $\dfrac{4}{9} \times \dfrac{2}{3}$

7. Multiply. Give your answers in the lowest terms (simplified) and as a mixed number, if possible.

a. $\dfrac{5}{12} \times \dfrac{2}{3}$	b. $9 \times \dfrac{4}{5}$

8. Find the area of a square with 1 7/8-inch sides.

9. After supper, a family of four had 1/3 of a pizza left.
   The next day, three people shared the remaining pizza equally.
   What fractional part of the *original* pizza did each of those people get?

10. **a.** How many 1/3-lb servings can you get from 3 pounds of chicken?

    b. Write a division sentence to match this situation.

11. Solve.

a. $\dfrac{1}{6} \div 3$	b. $6 \div \dfrac{1}{8}$	c. $\dfrac{9}{11} \div 3$

12. Draw a picture of some hearts, circles, and diamonds, so that 3/7 of
    the shapes are hearts, 2/7 of them are circles, and the rest are diamonds.
    What is the ratio of hearts to circles to diamonds?

13. A jar contains white and blue marbles in a ratio of 2:3. If there are 400 marbles in all,
    how many are white?

Chapter 8 Test

1. Classify each triangle by its sides and its angles. Name each quadrilateral.

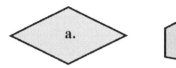

a. _____   b. _____

c. _____

d. _____

e. _____

2. A parallelogram's sides measure 5 in., 5 in., 5 in., and 5 in.
   Is it also a kite?   A trapezoid?   A square?

3. Answer.

   a. Is a square also a kite?
   Why or why not?

   b. Is a rhombus also a trapezoid?
   Why or why not?

   c. Can a rectangle sometimes be a kite?
   If yes, sketch one example.

   d. Could an equilateral triangle sometimes be a right triangle?
   If yes, sketch an example. If not, explain why not.

4. Which of these terms (perimeter, area, or volume) fits the situation; if you need to find out...

    a. how much fence is needed to go around a yard?

    b. how much water fits into a bottle?

    c. how big a carpet will cover the floor?

5. Draw an isosceles triangle with 30° base angles.
   You can choose its side lengths.
   What is the angle measure of its top angle?

6. A square has a perimeter of 4 inches. What is its area?

7. The dimensions of this box are 2 ft by 1.5 ft by 1.5 ft.
   What is its volume?

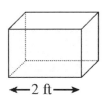

←2 ft→

8. A book measures 15 cm × 30 cm × 1.5 cm.
   You make a stack of six books.

    a. What is the volume of one book?

    b. What is the volume of the stack?

9. **a.** Plot the points, and connect them with line
     segments to form a triangle. Classify
     it by its angles and sides.

     Triangle: (0, 2), (0, 7), (4, 5)

     _____ and

     _____

    b. Draw a circle with center point (4, 4)
     and a radius of 3 units.

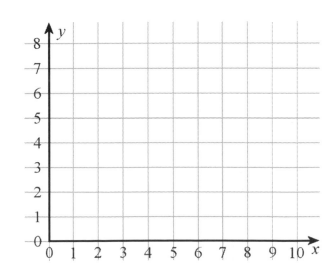

Math Mammoth Grade 5 End of the Year Test Instructions

This test is quite long, because it contains lots of questions on all of the major topics covered in *Math Mammoth Grade 5 Complete Curriculum*. Its main purpose is to be a diagnostic test—to find out what the student knows and does not know. The questions are quite basic and do not involve especially difficult word problems.

Since the test is so long, I do not recommend that you have your child/student do it in one sitting. Break it into 3-5 parts and administer them on consecutive days, or perhaps on morning/evening/morning/evening. Use your judgment.

A calculator is not allowed.

The test is evaluating the student's ability in the following content areas:

- the four operations with whole numbers
- the concept of an equation; solving simple equations
- divisibility and factoring
- place value and rounding with large numbers
- solving word problems, especially those that involve a fractional part of a quantity
- the concept of a decimal and decimal place value
- all four operations with decimals, to the hundredths
- coordinate grid, drawing a line graph, and finding the average
- fraction addition and subtraction
- equivalent fractions and simplifying fractions
- fraction multiplication
- division of fractions in special cases (a unit fraction divided by a whole number, and a whole number divided by a unit fraction)
- classifying triangles and quadrilaterals
- area and perimeter
- volume of rectangular prisms (boxes)

In order to continue with the *Math Mammoth Grade 6 Complete Worktext*, I recommend that the child gain a score of 80% on this test, and that the teacher or parent review with him any content areas in which he may be weak. The exception to this rule is integers, because they will be reviewed in detail in 6th grade. Children scoring between 70% and 80% may also continue with grade 6, depending on the types of errors (careless errors or not remembering something, versus a lack of understanding). Again, use your judgment.

Grading

My suggestion for points per item is as follows. The total is 171 points. A score of 137 points is 80%.

Question #	Max. points	Student score
The Four Operations		
1	2 points	
2	6 points	
3	2 points	
4	2 points	
5	2 points	
6	2 points	
7	3 points	
	subtotal	/ 19
Large Numbers		
8	2 points	
9	1 point	
10	1 point	
11	4 points	
	subtotal	/ 8
Problem Solving		
12	3 points	
13	3 points	
14	3 points	
15	3 points	
16	3 points	
17	3 points	
	subtotal	/ 18
Decimals		
18	4 points	
19	6 points	
20	3 points	
21	3 points	
22	3 points	
23	3 points	
24	9 points	
25	6 points	
26	9 points	
27	3 points	
28	3 points	
	subtotal	/52

Question #	Max. points	Student score
Graphs		
29	3 points	
30	2 points	
31	4 points	
	subtotal	/9
Fractions		
32	3 points	
33	4 points	
34	4 points	
35	2 points	
36	4 points	
37	2 points	
38	5 points	
39	3 points	
40	2 points	
41	4 points	
42	2 points	
43	2 points	
44	4 points	
	subtotal	/41
Geometry		
45	4 points	
46	4 points	
47	2 points	
48	3 points	
49	3 points	
50	3 points	
51	1 point	
52	4 points	
	subtotal	/24
	TOTAL	/171

Math Mammoth End-of-the-Year Test - Grade 5

The Four Operations

1. Solve (without a calculator).

   a. $1{,}035 \div 23$

   b. 492×832

2. Solve.

   a. $x - 56{,}409 = 240{,}021$

   b. $7{,}200 \div Y = 90$

   c. $N \div 14 = 236$

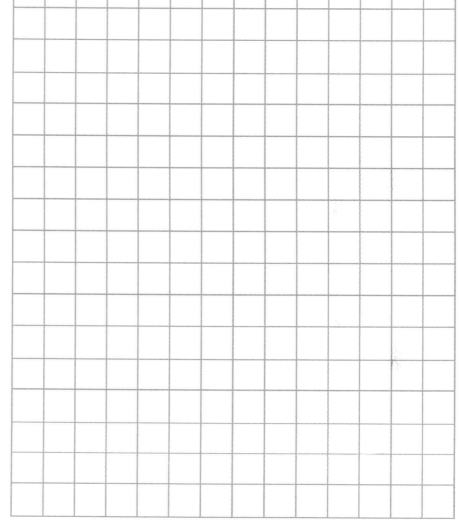

3. Write an equation to match this model, and solve it.

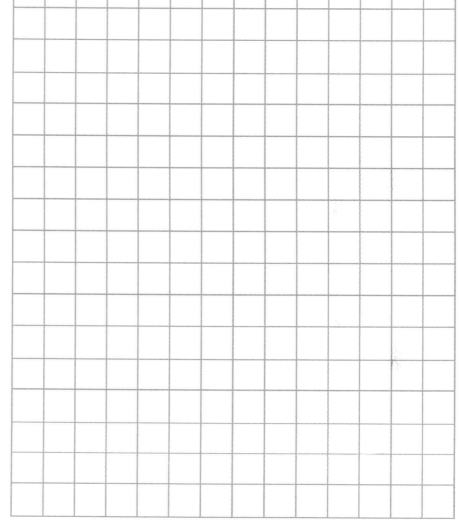

4. Place parentheses into the equations to make them true.

   a. $42 \times 10 = 10 - 4 \times 70$

   b. $143 = 13 \times 5 + 6$

5. Write a <u>single</u> expression (number sentence) for the problem, and solve.

A store was selling movies that originally cost $19.95 with a $5 discount.
Mia bought five of them. What was the total cost?

6. Is 991 divisible by 4?

   Why or why not?

7. Factor the following numbers to their prime factors.

a. 26 / \	**b.** 40 / \	**c.** 59 / \

Large Numbers

8. Write the numbers.

   a. 70 million 16 thousand 90

   b. 32 billion 232 thousand

9. Estimate the result of 31,933 × 305.

10. What is the value of the digit 8 in the number **56,782,010,000**?

11. Round these numbers to the nearest thousand, nearest ten thousand, nearest hundred thousand, and nearest million.

number	593,204	19,054,947
to the nearest 1,000		
to the nearest 10,000		
to the nearest 100,000		
to the nearest million		

Problem Solving

12. Jack has an 8-ft long board. He cuts off 1/6 of it.
    How long is the remaining piece, in feet and inches?

13. A website charges a fixed amount for each song download.
    If you can download six songs for $4.68, then how much would
    it cost to download ten songs?

14. A lunch in a fancy restaurant is three times as expensive as a lunch in a cafeteria.
    The lunch in the fancy restaurant costs $36. In a 5-day workweek, Mary eats at the
    fancy restaurant once, and in the cafeteria the rest of the days. How much does she
    spend on lunches in that week?

15. A blue swimsuit costs $42 and a red swimsuit costs 5/6 as much. How much would the two swimsuits cost together?

Mark the $42 in the bar model. Mark what is not known with "?". Solve.

16. A bag has green and purple marbles. Two-fifths of the marbles are green, and the rest are purple.

    a. Draw a bar model for this situation.

    b. If there are 134 green marbles, how many are purple?

17. Karen and Ann share the cost of a DVD that costs $29.90 so that Karen pays 3/5 of it and Ann pays 2/5 of it.

    a. *Estimate* how much each person will pay.

    b. Find the exact amount of how much each person will pay.

Decimals

18. Write the decimals indicated by the arrows.

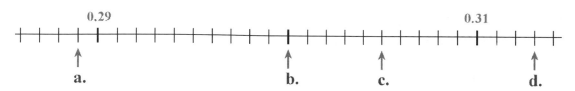

a. _____   b. _____   c. _____   d. _____

19. Complete.

a. $0.9 + 0.05 =$ _____	b. $0.28 +$ _____ $= 1$	c. $0.82 - 0.2 =$ _____
d. $1.3 - 0.04 =$ _____	e. $0.25 + 0.8 =$ _____	f. _____ $- 0.2 = 0.17$

20. Write as decimals.

a. $\dfrac{8}{100} =$ b. $\dfrac{81}{1000} =$ c. $5\dfrac{21}{100} =$

21. Write as fractions or mixed numbers.

a. 0.048 b. 1.004 c. 7.22

22. Compare, and write $<$ or $>$.

a. 0.31 ☐ 0.031 b. 0.43 ☐ 0.093 c. 1.6 ☐ 1.29

23. Round the numbers to the nearest one, nearest tenth, and nearest hundredth.

rounded to...	nearest one	nearest tenth	nearest hundredth	rounded to...	nearest one	nearest tenth	nearest hundredth
5.098				0.306			

24. Solve.

a. $0.4 \times 7 =$	d. $10 \times 0.05 =$	g. $1.1 \times 0.3 =$
b. $0.4 \times 0.7 =$	e. $100 \times 0.05 =$	h. $70 \times 0.9 =$
c. $0.4 \times 700 =$	f. $1000 \times 0.5 =$	i. $20 \times 0.09 =$

25. Divide.

a. $0.36 \div 6 =$	**c.** $3 \div 100 =$	**e.** $16 \div 10 =$
b. $5.6 \div 7 =$	**d.** $0.7 \div 10 =$	**f.** $71 \div 100 =$

26. Convert.

a. 0.2 m = _____ cm	**b.** 0.4 L = _____ ml	**c.** 56 oz = _____ lb _____ oz
37 cm = _____ m	3.5 kg = _____ g	74 in. = _____ ft _____ in.
2.9 km = _____ m	240 g = _____ kg	15 C = _____ qt _____ C

27. Two liters of ice cream is divided equally into nine bowls. Calculate how much ice cream is in **TWO** bowls, to the nearest milliliter.

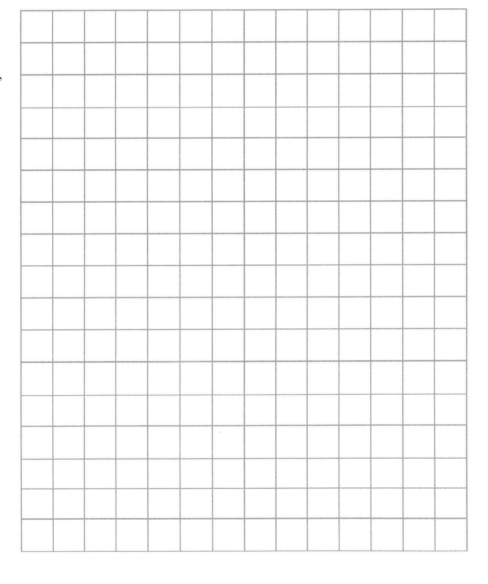

28. Calculate.

   a. $4.2 - 2.78$

   b. $71.40 \div 5$

   c. 2.2×6.4

29. Plot the points from the "number rule" on the coordinate grid.

    The rule for *x*-values:
    start at 0, and add 1 each time.

    The rule for *y*-values:
    start at 1, and add 2 each time.

x	0	1				
y	1					

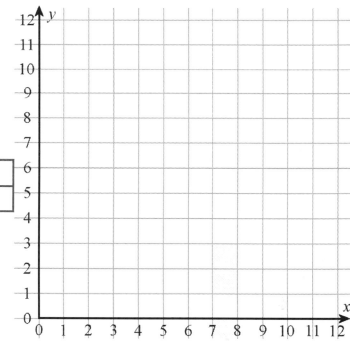

30. Draw in the grid a circle with a center point at (8, 4), and a radius of 3 units.

31. The table below gives the amount of sales in a grocery store from Monday through Friday.

Day	Sales (thousands of dollars)
Mon	125
Tue	114
Wed	118
Thu	130
Fri	158

    a. Make a line graph.

    b. Calculate the average daily sales in this period.

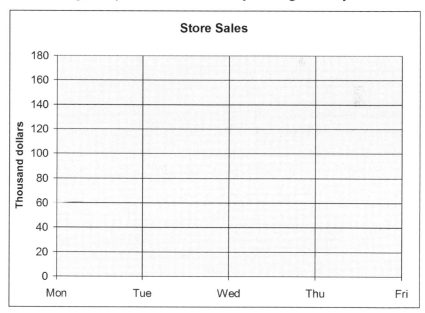

32. Add and subtract.

a.	b.	c. $3\frac{7}{10}$
$3\frac{7}{9}$ $+\ 2\frac{5}{9}$ _____	$5\frac{1}{6}$ $-\ 2\frac{5}{6}$ _____	$+\ 2\frac{8}{10}$ $+\ 7\frac{3}{10}$ _____

33. Mark the fractions on the number line. $\frac{3}{4}$, $\frac{1}{3}$, $\frac{4}{6}$, $\frac{5}{12}$

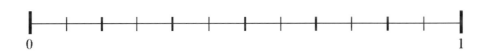

0   1

34. If you can find an equivalent fraction, write it. If you cannot, cross the whole problem out.

a. $\frac{5}{6} = \frac{}{20}$	b. $\frac{2}{7} = \frac{}{28}$	c. $\frac{3}{8} = \frac{15}{}$	d. $\frac{2}{9} = \frac{6}{}$

35. Find the errors in Mia's calculation and correct them.

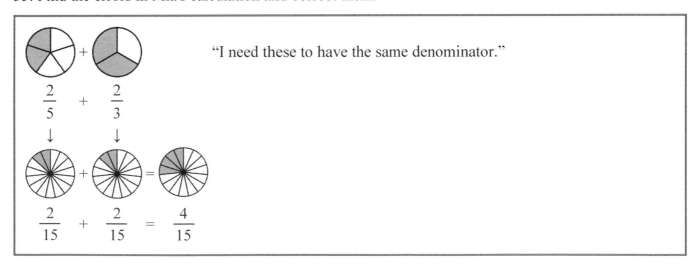

"I need these to have the same denominator."

$$\frac{2}{5} + \frac{2}{3}$$

$$\frac{2}{15} + \frac{2}{15} = \frac{4}{15}$$

36. Add and subtract the fractions and mixed numbers.

a. $\dfrac{1}{3} + \dfrac{5}{6}$	**b.** $\dfrac{4}{5} - \dfrac{1}{3}$
c. $6\dfrac{1}{8} - \dfrac{1}{2}$	**d.** $6\dfrac{7}{9} + 3\dfrac{1}{2}$

37. You need 2 3/4 cups of flour for one batch of rolls.
    Find how much flour you would need for three batches of rolls.

38. Compare the fractions, and write $<$, $>$, or $=$ in the box.

a. $\dfrac{6}{9} \ \square \ \dfrac{6}{13}$ **b.** $\dfrac{6}{13} \ \square \ \dfrac{1}{2}$ **c.** $\dfrac{5}{10} \ \square \ \dfrac{48}{100}$ **d.** $\dfrac{1}{4} \ \square \ \dfrac{25}{100}$ **e.** $\dfrac{5}{7} \ \square \ \dfrac{7}{10}$

39. Simplify the following fractions if possible. Give your answer as a mixed number when you can.

a. $\dfrac{21}{15} =$	**b.** $\dfrac{29}{36} =$	**c.** $\dfrac{42}{48} =$

40. Is the following multiplication correct?
    If not, correct it. $\dfrac{2}{3} \times$ $=$

41. Multiply the fractions, and shade a picture to illustrate the multiplication.

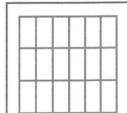 **a.** $\dfrac{1}{3} \times \dfrac{5}{6}$

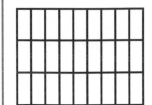 **b.** $\dfrac{2}{9} \times \dfrac{2}{3}$

42. How many 1/4-inch pieces can you cut
from a string that is 15 inches long?

43. Three people share half a pizza evenly. What fractional
part of the original pizza does each one get?

44. Solve. Give your answer as a mixed number and in a simplified form.

a. $\dfrac{7}{6} \times 9$	**b.** $\dfrac{1}{7} \div 3$
c. $\dfrac{4}{5} \times 3\dfrac{2}{3}$	**d.** $2 \div \dfrac{1}{9}$

45. Measure the sides of the triangle in inches. Find its perimeter.

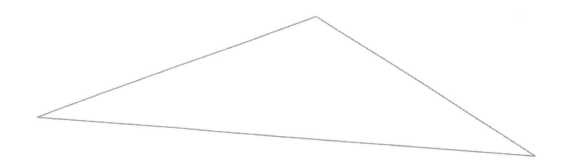

46. Below you see two triangles and two quadrilaterals. Classify the triangles according to their sides and angles. Name the quadrilaterals.

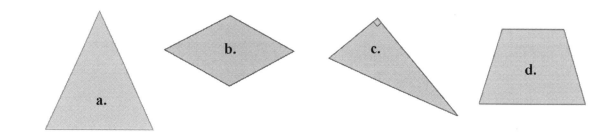

a. _____

b. _____

c. _____

d. _____

47. **a.** A square has a perimeter of 12 m. What is its area?

b. A square has an area of 25 ft^2. What is its perimeter?

48. Is a square a trapezoid? Why or why not?

49. Can an obtuse triangle be isosceles?
    If not, explain why not.
    If yes, sketch an example.

50. **a.** Draw a right triangle with 5 cm and 7 cm perpendicular sides.

    b. Find its perimeter.

    c. Measure its angles. They measure _____°, _____°, and _____°.

51. This is a rectangular prism.
    Find its volume.

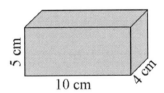

5 cm

10 cm

4 cm

52. Matthew has a rainwater collection tank in his yard that is rectangular,
    like a box. It is 1.2 m long, 60 cm wide, and 1 m tall.

    a. Find the volume of the tank in cubic <u>meters</u>.

    b. One morning, after a rainy night, the tank is about 1/3 full.
    About how many liters of water are in the tank?
    1 cubic meter equals 1,000 liters.

USING THE CUMULATIVE REVIEWS

The cumulative reviews include problems from various chapters of the Math Mammoth complete curriculum, up to the chapter named in the review. For example, a cumulative review for chapters 1-6 may include problems from chapters 1, 2, 3, 4, 5, and 6. The review for chapters 1-6 can be used any time after chapter 6 has been studied.

I am providing these reviews as optional, additional practice and review for the student. This means that the student does not necessarily HAVE TO do all these review problems. The teacher should decide when, if ever, these reviews are used.

I would recommend using at least 2 or 3 of these review files during the school year. The teacher can also use the reviews in a diagnostic way to find out what kinds of problems the student has trouble with.

Math Mammoth complete curriculum also comes with an easy worksheet maker, which is the perfect tool to make lots of problems of a specific type, especially when it comes to calculation skills.

You can access the worksheet maker online at

http://www.homeschoolmath.net/worksheets/grade_5.php

So, if you find that the student has trouble with some specific type of a problem while doing a cumulative review, use the worksheet maker to make more problems of that type (if available; the worksheet maker does not make all kinds of problems, for example word problems).

Cumulative Review, Grade 5, Chapters 1-2

1. Place parentheses into these equations to make them true.

| a. $90 + 70 + 80 \times 2 = 390$ | b. $378 = 6 \times 8 + 13 \times 3$ | c. $90 \times 4 = 180 - 60 \times 3$ |

2. Draw a bar model to illustrate the equations. Then solve the equations.

| a. $4x + 120 = 200$ | b. $25 + 3x = 52$ |
| | |

3. Divide. Use the space on the left for building a multiplication table of the divisor. Lastly, check.

$2 \times 15 = 30$	a. $15 \overline{)9450}$	$\times\ 1\ 5$
	b. $14 \overline{)4508}$	$\times\ 1\ 4$

4. Solve the word problems.

   a. Jim earned a total of $1,920 dollars in four weeks.
      How much did he earn in one week?

   b. Joe entered his sled and dogs in 11 races last year.
      The races were all held on the same 136-mile race course.
      How many miles total did Joe and his dogs race last year?

5. Which expression(s) match the problem? Also, solve the problem.

Greg bought four flashlights for $9 each, and paid with $50. What was his change?	**(1)** $50 − $9 + $9 + $9 + $9 **(2)** $50 − ($9 − $9 − $9 − $9) **(3)** $50 − ($9 + $9 + $9 + $9)	**(4)** 4 × $9 − $50 **(5)** $50 − 4 × $9 **(6)** $50 + 4 − $9

6. First, estimate the answer to the multiplication problem. Then multiply.

a. Estimate: _____	**b.** Estimate: _____	**c.** Estimate: _____
_____	_____	_____
$$\begin{array}{r} 1\ 7\ 3 \\ \times\ \ \ 3\ 5 \\ \hline \end{array}$$	$$\begin{array}{r} 2\ 6\ 9 \\ \times\ 5\ 3\ 7 \\ \hline \end{array}$$	$$\begin{array}{r} 8\ 9\ 2 \\ \times\ 3\ 4\ 0 \\ \hline \end{array}$$

Cumulative Review, Grade 5, Chapters 1-3

1. Divide. Use the space on the right for building a multiplication table of the divisor. Then check.

$2 \times 21 = 42$	$21\overline{)8\ 1\ 6\ 9}$	$\times\ 2\ 1$

2. Solve in the right order. First, you can enclose the operation to be done in a "bubble" or a "cloud."

a. $94 + 12 \times 5 \div 2 = $ _____	**b.** $(22 - 9) \times 2 + 58 = $ _____
c. $43 + (55 + 5) \div 5 = $ _____	**d.** $700 - 30 \times (3 + 4) = $ _____

3. Solve mentally.

a. $43 - 17 = $ _____ $71 - 43 = $ _____	**b.** $54 - 19 + 12 = $ _____ $85 - 25 + 75 = $ _____	**c.** $1{,}200 - $ _____ $= 750$ $2{,}000 - 800 - $ _____ $= 600$

4. Write the numbers.

   a. 78 billion   38   16 thousand

   b. 844 billion   12 million   704

5. Round these numbers to the nearest thousand, nearest ten thousand, nearest hundred thousand, and nearest million.

number	32,274,302	64,321,973	388,491,562	2,506,811,739
to the nearest 1,000				
to the nearest 10,000				
to the nearest 100,000				
to the nearest million				

6. Complete the addition path using mental math.

43,199,000	add 10,000 →		add a million →	

add ↓ 100 thousand

	add 10 million ←		add a thousand ←	

7. Write an expression to match each written sentence.

a. The product of 5 and 6 is added to 50.	**b.** The difference of 9 and 6 is subtracted from 10.

8. Write a single expression using numbers and operations for the problem, not just the answer!

A teacher bought 21 notebooks for $2 each, 20 rulers for $1.50 each, and chalk for $12.
What was the total cost?

9. Add.

a. 521,607,090 + 4,293,991,092	**b.** 77,630,087 + 884,000,299 + 84,926,571

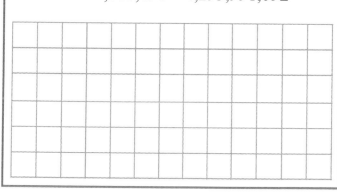
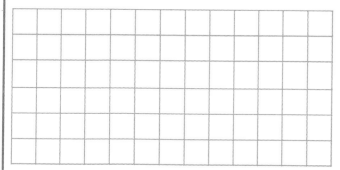

10. Estimate first, using mental math. Then find the exact answer and the error of your estimation using a calculator.

a. 2,933 × 213	**b.** 152 × 89 × 7,932
My estimation: _____	My estimation: _____
Exact answer: _____	Exact answer: _____
Error of estimation: _____	Error of estimation: _____

Cumulative Review, Grade 5, Chapters 1-4

1. Jake earned $125 and his sister earned
   4/5 as much. How much did Jake and
   his sister earn together?

   Mark the information in the bar model, and solve.

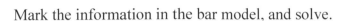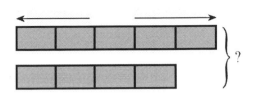

2. Mom is buying a thermometer, and the store has two kinds.
   The pricier one costs $10.40, and the cheaper just 3/4 as much.

   How much more does the expensive thermometer cost than the cheaper one?

3. Helen has 120 marbles and Julie has 2/5 as many.
   How many more marbles does Helen have than Julie?

4. Ann is an English teacher. She has 150 students in her English classes this year,
   and 6/50 of them were not in her classes last year.

   a. How many new students does she have?

   b. Out of the new students, 1/3 have never studied English before.
   How many of the new students have studied English before?

5. Divide. Use the space on the right for building a multiplication table of the divisor. Then check.

$2 \times 37 = 74$	$37\overline{)6\ 7\ 3\ 4}$	$\times\ 3\ 7$

6. Solve for the unknown N or M.

a. $4 \times M = 200$	**b.** $M \div 6 = 8$	**c.** $4{,}500 \div M = 50$
d. $7 \times N = 56{,}000$	**e.** $N \div 30 = 700$	**f.** $48{,}000 \div N = 600$

7. Write an expression to match each written sentence.

a. The quotient of 350 and x equals 5.	**b.** The difference of 15 and 6 is added to 8.

8. Find a number to fit in the box so the equation is true.

a. $36 = (\boxed{} + 9) \times 3$	**b.** $7 \times 7 = 4 \times \boxed{} + 5$	**c.** $19 = (84 \div \boxed{}) - 2$

9. Round these numbers to the nearest thousand, nearest ten thousand, nearest hundred thousand, and nearest million.

number	97,302	25,096,199	709,383,121	89,534,890,066
to the nearest 1,000				
to the nearest 10,000				
to the nearest 100,000				
to the nearest million				

Cumulative Review, Grade 5, Chapters 1-5

1. Round the numbers to the nearest unit (one), to the nearest tenth, and to the nearest hundredth.

Round this to the nearest →	unit (one)	tenth	hundredth
4.925			
6.469			

Round this to the nearest →	unit (one)	tenth	hundredth
5.992			
9.809			

2. Jake worked for 56 days on a farm, and Ed worked for 14 days less.
   How many days did the two boys work together?

3. Add using mental math

a. $0.3 + 0.07 = $ _____	**b.** $0.19 + 0.002 = $ _____	**c.** $0.028 + 0.3 = $ _____
d. $1.05 + 0.4 = $ _____	**e.** $0.49 + 0.56 = $ _____	**f.** $0.006 + 0.5 = $ _____

4. Multiply both the dividend and the divisor by 10, repeatedly, until you get a *whole-number divisor*.
   Then, divide using long division.

a. $0.927 \div 0.3$ Check: _____)	**b.** $0.646 \div 0.08$ Check: _____)

5. Convert the measuring units.

a. 0.5 m = _____ cm	b. 4.2 L = _____ mL	c. 800 g = _____ kg
0.06 m = _____ cm	400 mL = _____ L	4,550 m = _____ km
2.2 km = _____ m	5,400 g = _____ kg	2.88 kg = _____ g

6. Jerry bought three packages of AA batteries and six packages of AAA batteries.
   The total was $17.04. One package of the AA batteries cost $1.88.
   What does one package of AAA batteries cost?

7. Divide in two ways: first by indicating a remainder, then by long division. Give your answers to two
   decimal digits.

a. 31 ÷ 6 = _____ R _____ Check:	b. 43 ÷ 4 = _____ R _____ Check:

8. First, estimate the answer. Then, multiply. Do not forget the decimal points.

a. 2.09 × 11.5	**b.** 73 × 2.14	**c.** 7.1 × 3.02
Estimate: _____	Estimate: _____	Estimate: _____
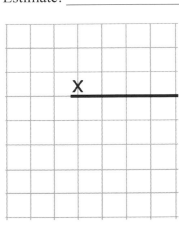	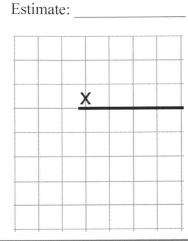	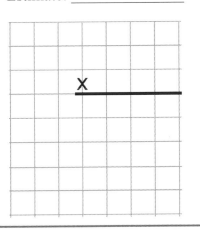

9. Alex bought seven packets of cucumber seeds for $13.23 total. He also bought seven flower plants that were originally $3.20 each but the price was reduced by 4/10.

   a. What did one packet of seeds cost?

   b. How much did one flower plant cost?

   c. What was the total cost?

10. Multiply and divide.

a. 10 × 0.07 = _____	**b.** 100 × 0.63 = _____	**c.** 1000 × 0.029 = _____
d. 0.8 ÷ 10 = _____	**e.** 4.5 ÷ 100 = _____	**f.** 76 ÷ 1000 = _____

Cumulative Review, Grade 5, Chapters 1-6

1. First *estimate* the answer to each multiplication. Then multiply to find out the exact answer.

a. 290×277	**b.** 525×416	**c.** 897×186
Estimate: _____	Estimate: _____	Estimate: _____
	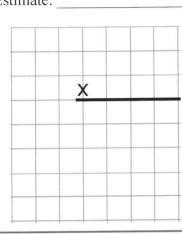	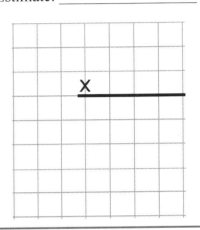

2. While jogging, Rebecca saw a big snake on the path 250 m before the end of the 2.4-km jogging track. She got so scared that she turned back on the track and jogged back to the beginning of the track. Find the total distance that she jogged on the track.

3. Angi and Rebekkah split their total earnings of $100 so that Angi got $10 more than Rebekkah. How much did each one get?

4. Find the missing factor.

a. $10 \times$ _____ $= 4.0$	**b.** $5 \times$ _____ $= 6.0$	**c.** _____ $\times 0.11 = 3.3$
d. _____ $\times 0.3 = 0.06$	**e.** $2 \times$ _____ $\times 1.2 = 48$	**f.** $3 \times$ _____ $\times 0.5 = 6$

5. Solve the equations.

a. $y - 0.57 = 1.1$	b. $7.319 + z = 9$

6. Calculate the average (the mean) of the data set. Do not use a calculator.

    21, 19, 25, 22, 13, 15, 24, 12, 11

7. Write in expanded form.

   a. 0.908

   b. 543.2

8. Divide. Mental math will work!

a. $0.8 \div 2 =$ _____	b. $0.36 \div 6 =$ _____	c. $0.25 \div 0.05 =$ _____
d. $0.16 \div 4 =$ _____	e. $0.54 \div 0.06 =$ _____	f. $1 \div 0.05 =$ _____

9. A group of 37 medical students traveled through ten states
   to view new technology in some progressive hospitals.
   They had to share the expense of $99,000 for the trip.
   What was each student's share of the expenses?
   Round your answer to the nearest dollar.

10. Ashley bought a 1/2-gallon carton of milk, and used 2 cups
    of it for baking. How many *cups* of milk are left?

11. Ava is 4 ft 8 in. tall and Eva is 61 inches tall.
    Who is taller? How many inches taller?

12. Juan is mailing 36 CDs that weigh 6 ounces each.
    What is the total weight of the CDs, in pounds and ounces?

13. The following numbers describe the distance in kilometers that 16 employees of a small
    company drive to work.   15 7 22 6 16 25 31 45 7 11 9 19 25 4 15 18

    a. Fill in the frequency table below. Make four or five categories. Then draw a histogram.

    b. Find the average number of kilometers the employees drive to work.

    c. Find the mode of the data set.

distance	frequency

Cumulative Review, Grade 5, Chapters 1-7

1. Solve in the right order!

a.	b.	c.
$13 \times 4 + 18 = $ _____	$(2 + 60 \div 4) \times 3 = $ _____	$10 \times (9 + 18) \div 3 = $ _____
$4 + 8 \div 8 = $ _____	$2 + 30 \times (7 + 8) = $ _____	$5 \times (200 - 190 + 40) = $ _____

2. Joe bought 100 apples for $0.23 each. He divided them equally into ten small bags.

   a. What was the total cost for 100 apples?

   b. What was the value of each small bag of apples?

3. Compare the fractions.

   a. $\dfrac{2}{3} \square \dfrac{5}{8}$　　　**b.** $\dfrac{1}{4} \square \dfrac{4}{9}$　　　**c.** $\dfrac{5}{6} \square \dfrac{5}{7}$　　　**d.** $\dfrac{6}{8} \square \dfrac{3}{4}$

4. In what place is the underlined digit? What is its value?

a. 791,4<u>5</u>6,030	**b.** 2,094,806,391
Place: _____	Place: _____
Value: _____	Value: _____

5. Make a line graph of this data for the Oak Bend Hospital.

Year	Babies Born
1950	225
1960	340
1970	460
1980	525
1990	580
2000	520
2010	490

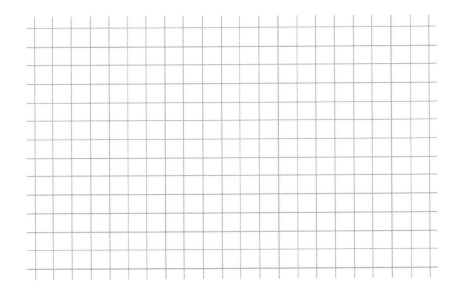

6. Write an equation to match the bar model. Then, solve for x.

| x | x | x | x | 176 |

$\longleftarrow$ —————— 516 —————— $\longrightarrow$

7. A hotel maintains two jogging paths in the woods. The shorter one is 1.2 km long and the other is four times as long. If you jog both paths, then what is the total distance you have jogged?

8. Shelly is going to buy eight pounds of oranges for $1.19 a pound, and six pounds of bananas for $0.88 a pound.

   a. Estimate the total cost to the nearest dollar.

   b. The cashier announces that Shelly will get 1/5 off of her bill for being a loyal customer. Now calculate what Shelly pays. Use the actual total in your calculation, not the rounded total.

9. Add and subtract.

a. $6\frac{6}{11} - 3\frac{2}{5}$	**b.** $6\frac{6}{7} + 1\frac{1}{2}$
c. $7\frac{9}{10} - 1\frac{1}{4}$	**d.** $3\frac{2}{5} + 2\frac{5}{6}$

10. These decimal divisions are not even. Round the answers to the nearest hundredth.

a. $3.377 \div 3$	b. $22.91 \div 11$	c. $62.6 \div 7$
$)\overline{}$	$)\overline{}$	$)\overline{}$

11. Divide in two ways: first by indicating a remainder, then by long division. Give your answers to two decimal digits.

a. $31 \div 6 =$ _____ R _____	b. $43 \div 4 =$ _____ R _____
$)\overline{}$ Check:	$)\overline{}$ Check:

Cumulative Review, Grade 5, Chapters 1-8

1. Draw the hearts and diamonds as stated in the ratio in the box on the far right and fill in the fractions.

Picture or Diagram	As Fractions	As a Ratio
	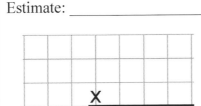 ——— of the shapes are hearts. ——— of the shapes are diamonds.	The ratio of hearts to diamonds is 1:5.

2. **a.** What is the ratio of circles to squares?

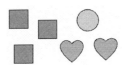

   b. What is the ratio of squares to all shapes?

3. Jenny made gingersnaps, chocolate drops, and oatmeal cookies in the ratio of 1 : 3 : 2. She made 72 cookies in total. How many were oatmeal cookies?

4. Find the volume of a box that is 3 inches deep, 5 1/2 inches wide, and 2 inches tall.

5. Solve by multiplying in columns.

a. 21.7×3.9	**b.** 0.52×0.8	**c.** 141×5.22
Estimate: _____	Estimate: _____	Estimate: _____
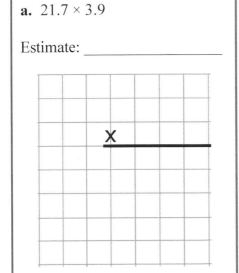		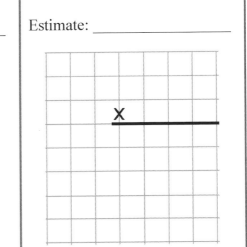

6. Scott is a plumber, and each day he has to drive around town to the clients' homes.
   The following numbers show how many kilometers Scott drove on ten different workdays:

   128 68 73 163 93 102 68 85 90 45

   a. Find the average number of kilometers
   Scott drove per day.

   b. Based on the average, calculate *approximately*
   how many kilometers Scott would drive at work
   in a year's time. Assume that he works 40 weeks
   a year, 5 days a week.

7. Name the following
   quadrilaterals.

   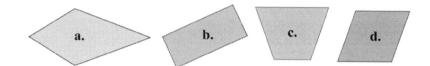

   a. _____

   b. _____

   c. _____ **d.** _____

8. Draw an isosceles right triangle
   with two 6-cm sides.

9. Multiply. Give your answers in the lowest terms (simplified) and as a mixed number, if possible.

a. $\dfrac{6}{8} \times \dfrac{2}{9}$	**b.** $\dfrac{9}{11} \times 2\dfrac{1}{3}$

10. A drinking glass measures 3/10 of a liter.
    How many glasses full of water do you get from a 3-liter pitcher?

11. **a.** Fill in the table how much weight Greg gained during each year.

    b. At what ages did he gain weight the fastest?

    c. How can you see these fast growth periods on the graph?

AGE (yrs)	WEIGHT (kg)	Weight gain from previous year
0	3.3 kg	-
1	10.2 kg	6.9 kg
2	12.3 kg	2.1 kg
3	14.6 kg	
4	16.7 kg	
5	18.7 kg	
6	20.7 kg	
7	22.9 kg	
8	25.3 kg	
9	28.1 kg	

AGE (yrs)	WEIGHT (kg)	Weight gain from previous year
10	31.4 kg	
11	32.4 kg	
12	37.0 kg	
13	40.9 kg	
14	47.0 kg	
15	52.6 kg	
16	58.0 kg	
17	62.7 kg	
18	65.0 kg	

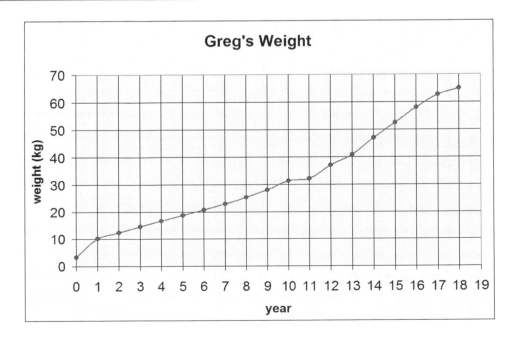

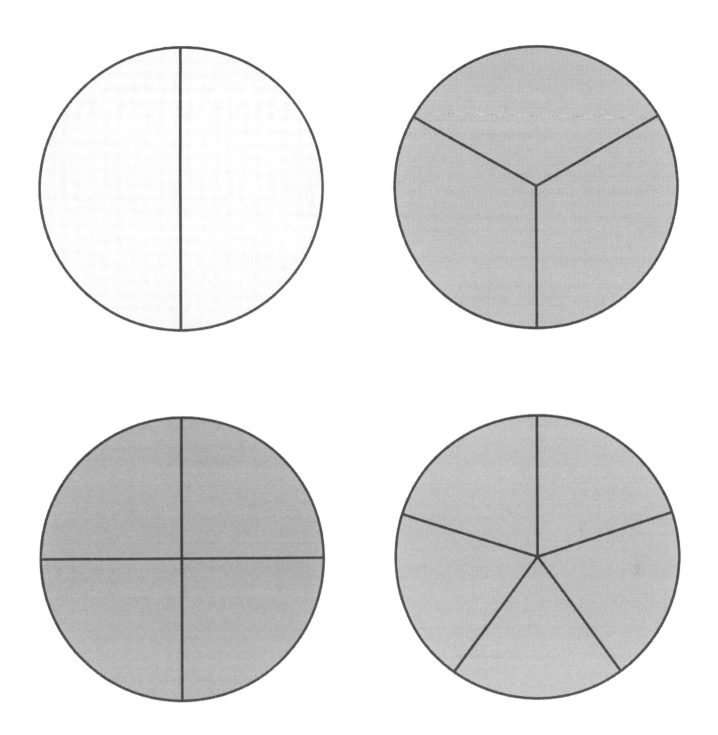

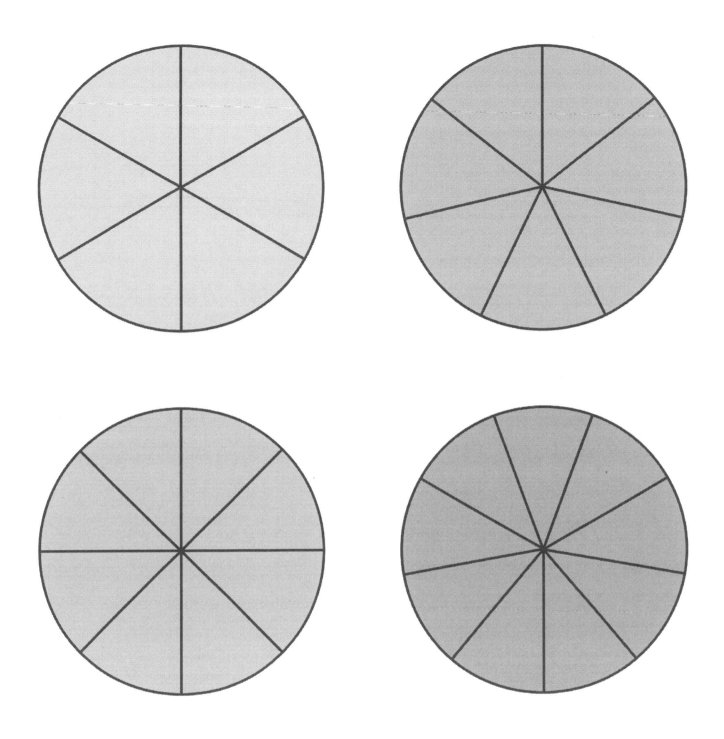

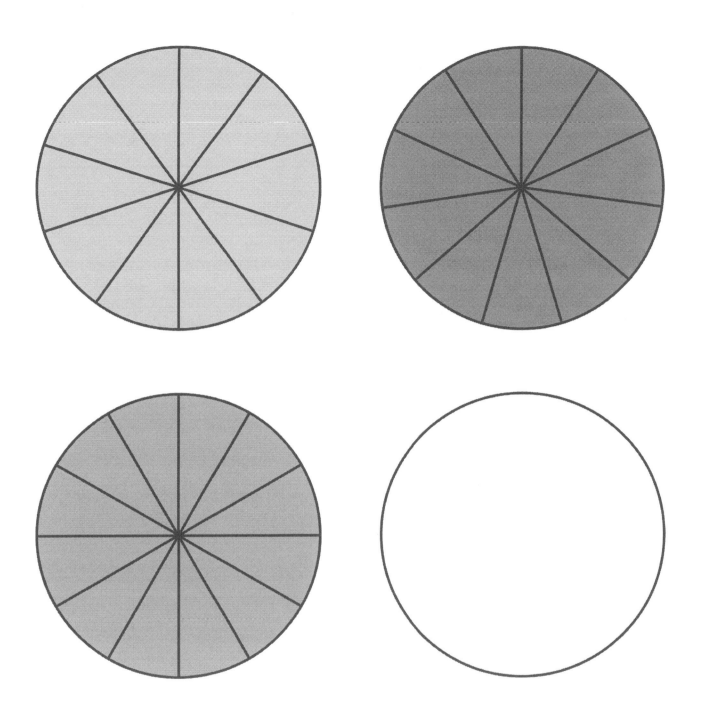

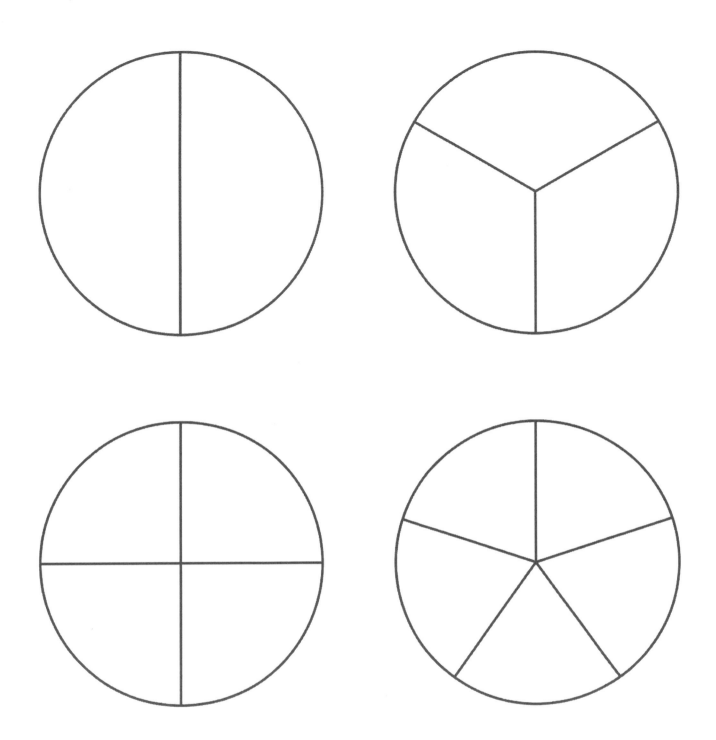

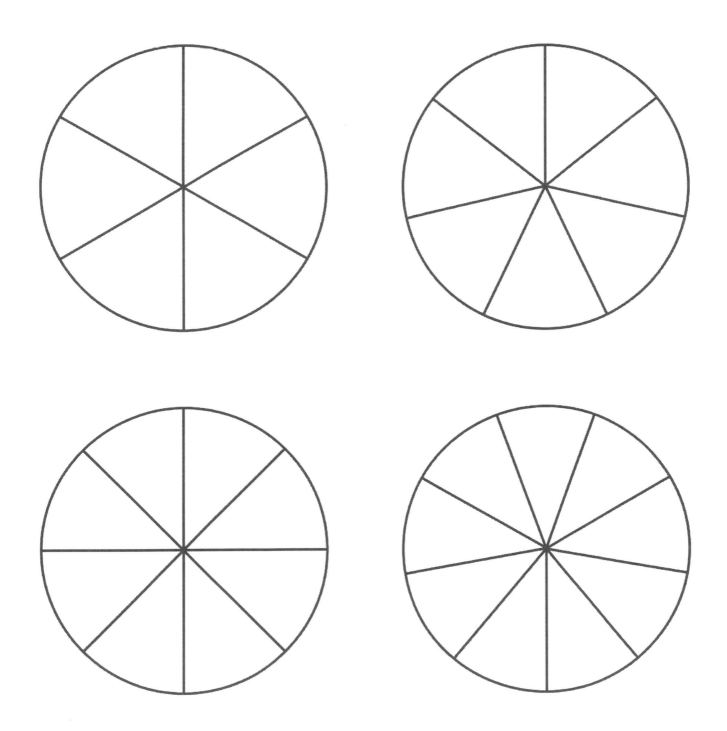

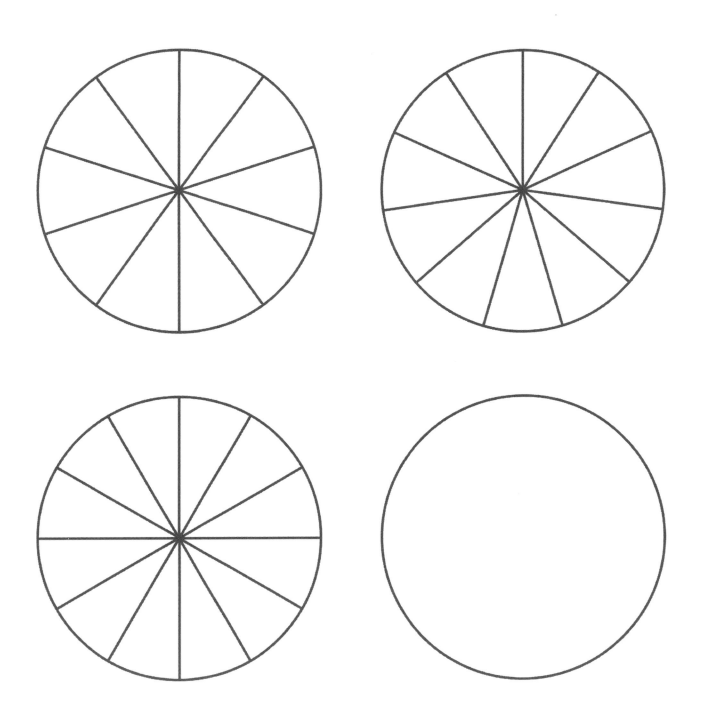

Made in the USA
Charleston, SC
20 December 2013